YOUR KNOWLEDGE HAS VALUE

- We will publish your bachelor's and master's thesis, essays and papers

- Your own eBook and book - sold worldwide in all relevant shops

- Earn money with each sale

Upload your text at www.GRIN.com and publish for free

T.S. Amar Anand Rao

Redox Electricity from Microbes to power LEDs

GRIN Verlag

Bibliografische Information der Deutschen Nationalbibliothek:

Die Deutsche Bibliothek verzeichnet diese Publikation in der Deutschen National-
bibliografie; detaillierte bibliografische Daten sind im Internet über http://dnb.d-
nb.de/ abrufbar.

Imprint:

Copyright © 2011 GRIN Verlag GmbH
Druck und Bindung: Books on Demand GmbH, Norderstedt Germany
ISBN: 978-3-656-09019-9

This book at GRIN:

http://www.grin.com/en/e-book/184277/redox-electricity-from-microbes-to-power-
leds

GRIN - Your knowledge has value

Der GRIN Verlag publiziert seit 1998 wissenschaftliche Arbeiten von Studenten, Hochschullehrern und anderen Akademikern als eBook und gedrucktes Buch. Die Verlagswebsite www.grin.com ist die ideale Plattform zur Veröffentlichung von Hausarbeiten, Abschlussarbeiten, wissenschaftlichen Aufsätzen, Dissertationen und Fachbüchern.

Visit us on the internet:

http://www.grin.com/

http://www.facebook.com/grincom

http://www.twitter.com/grin_com

Redox Electricity from Microbes to power LEDs

T.S. Amar Anand Rao

Molecular Biophysics Unit, Indian Institute of Science

Bangalore – 560012, Karnataka, India.

Abstract

Biogeochemical cycling through microbial succession in Winogradsky columns generates electricity. Here we use all electricals tools possible to harness this electricity and power Light Emitting Diodes for a year or more. The setup, biochemical mechanisms are analysed.

Introduction

Do the columns generate energy? Would a small light bulb light? Why might this occur? Where does the energy come from for the bacteria? (Woodrow Wilson Summer Biology Institute Biodiversity, 2000)

Just as the microbes in the Winogradsky column are changing their environment in ways that we can see, they are also making changes which are not so easily noticed. As the iron in the bottle is being used by the bacteria, there are processes at work changing the iron itself. Fe^{2+} is being oxidized to form Fe^{3+}, and Fe^{3+} is being reduced to form Fe^{2+}. In the simplest of terms, the difference between these two ions is a difference of electrons. As Fe^{2+} is being oxidized to form Fe^{3+}, an electron is lost from that atom of iron. It is the bacteria in the bottle which are helping this to happen. Some of the bacteria remove electrons from iron and transfer them to oxygen (an electron acceptor). Others act take the electrons from the food that was added and transfer them to the iron (an electron donor). This causes an imbalance in the number of free electrons at different heights in the bottle. Because we can not see these changes we must find another way to measure the changes taking place within the bottle. One way that we can measure the changes is by using a voltage meter and an ammeter. The voltage meter allows us to quantitatively measure the difference in the number of free electrons at the top of the bottle as compared to the bottom of the bottle (the electric potential). The ammeter measures the current flowing through the bottle. While working on this activity, remember the steps of the iron cycle and the conditions which are necessary for bacteria to interact with the iron. (National Science Foundation)

The pH-value played a crucial role for the development and current production of anodic microbial electroactive biofilms. It was demonstrated that only a narrow pH-window, ranging from pH 6 to 9, was suitable for growth and operation of biofilms derived from pH-neutral wastewater. Any stronger deviation from pH neutral conditions led to a substantial decrease in the biofilm performance. Thus, average current densities of 151, 821 and 730 $\mu A\ cm^{-2}$ were measured for anode biofilms grown and operated at pH 6, 7 and 9 respectively. The microbial diversity of the anode chamber community during the biofilm selection process was studied using the low cost method flow-cytometry. Thereby, it was demonstrated that the pH value as well as the microbial inocula had an impact on the resulting anode community structure. As shown by cyclic voltammetry the electron transfer thermodynamics of the biofilms was strongly depending on the solution's pH-value. (Susann Müller et al., 2011)

" Microbial Fuel Cells (MFC) work because some of the bacteria (exoelectrogens) found in creek and marine sediment can produce free electrons and these can generate electricity. The microbial activity in my fuel cell continued to be much more significant and interesting than in my spring Winogradsky column. The fuel cell was created using my most active Winogradsky column from last fall. The entire process has been interesting and I would definately do this with a class. Even smaller children enjoy it, as my son, who is in 2nd grade, helped me with the measurements and was able to understand the basic idea that tiny organisms in the mud were creating small amounts of electric current.

I used the methods described by Josh McCready and Tess Edmonds at the Geobacter Website: http://www.emunsing.com/portfolio/portfolio_content/44_designLab/MFC_Report.pdf to build my fuel cell with some minor adjustments due to materials availability.It is important to note that there are many different configurations and materials employed in constructing an MFC and new methods are being developed all over the world. For the purposes of this class and any use for younger children, I thought it would be best to follow a simple design with the fewest and simplest materials. I therefore followed the idea from the Geobactor site of constructing a sediment MFC, which simply uses sediment as a biomass and bacteria source (anodechamber), the overlying water as a substrate (cathode chamber), and the finer top sediment as a membrane to separate these two 'chambers'. 1. gradual increased voltage over time as the bacteria populate the electrode, 2. I did not take measurements within any 24 hour period, so I can not varify that the measurements would have oscillated during a 24 hour period as McCready and Edmonds suggest they should (indicating that the microbes go through natural daily cycles of activity), finally, they suggest completely sterilizing the MFC to prove that it will go to zero, validating that the microbes were generating electrical current while alive. I did not do this either." (http://My Winogradsky Column)

Materials and Methods

Procedure of making a Winogradsky column (Anderson et al 1999) the soil sample was cleaned of debris, stones, pebbles, grass clippings, leaves and moving insects. This is used as the control column and standard reference

In column, the electrochemical gradient potential between the top and the bottom of the standard column was monitored by inserting multimeter probes (Anderson, 1999)

Orpat India Multimeter to measure in mV or Volts was purchased.

We used to carbon rods obtained from old discarded AA size batteries as the electrodes, one at the top and one at the bottom of the column. Then using water resistant glue and washers we fixed the top and bottom of the empty bottle with these electrodes. Wirings were done to extend the electrodes to the other setup. (Figure 1)

The bottom portion was reserved for biomass for providing the electrochemical gradient by mixing the basic chemicals with red ferrous mud and filled the rest of the column with normal mud obtained form the pond.

We measured the current in the voltmeter from 0 day onwards for a year.

The idea was to charge supercapacitors connected with LEDs through a resistor (Figure 4)

First a super capacitor was found in old VCR, this cap is use for memory for clock.. Next part is led diode (any color) and resistor from 220 to 470ohm. Super Capacitor with 100 000 micro Farads is conected to LED with resistor, and the capacitor is charged it with DC power (www.instructable.com/ Supercapacitor) (Figure 3)

We charged this capacitor from electrodes from the Winogradsky column

Results

The electrochemical potential probed between the top and bottom of a column plate showed increase in value from 0 to 500mV with time from 0 to 90 days indicating the ability to produce tiny amounts of electricity that may be used to light up diode bulbs (Figure 2).

The carbon electrodes are non-corrosive. So a constant supply of current was obtained.

It fluctuated between 200mV to 500mV for over a year. After that peroid, it comes down. So we made use of the idea to charge supercapacitors connected with LEDs through resistor which stabilized the power conumption. The excess current is recharged into the capacitor and excess powers pass through the resistor to stably power the LED bulb which needs 1000mV.

Figure Legends

Figure 1 : Electrodes for the winogradsky bottle

figure 2: Voltmeter reading the electrochemical potential

Figure 3 : Circuit of capacitor, resistor and LED

Figure 4: Rechargable LED

Figure 1

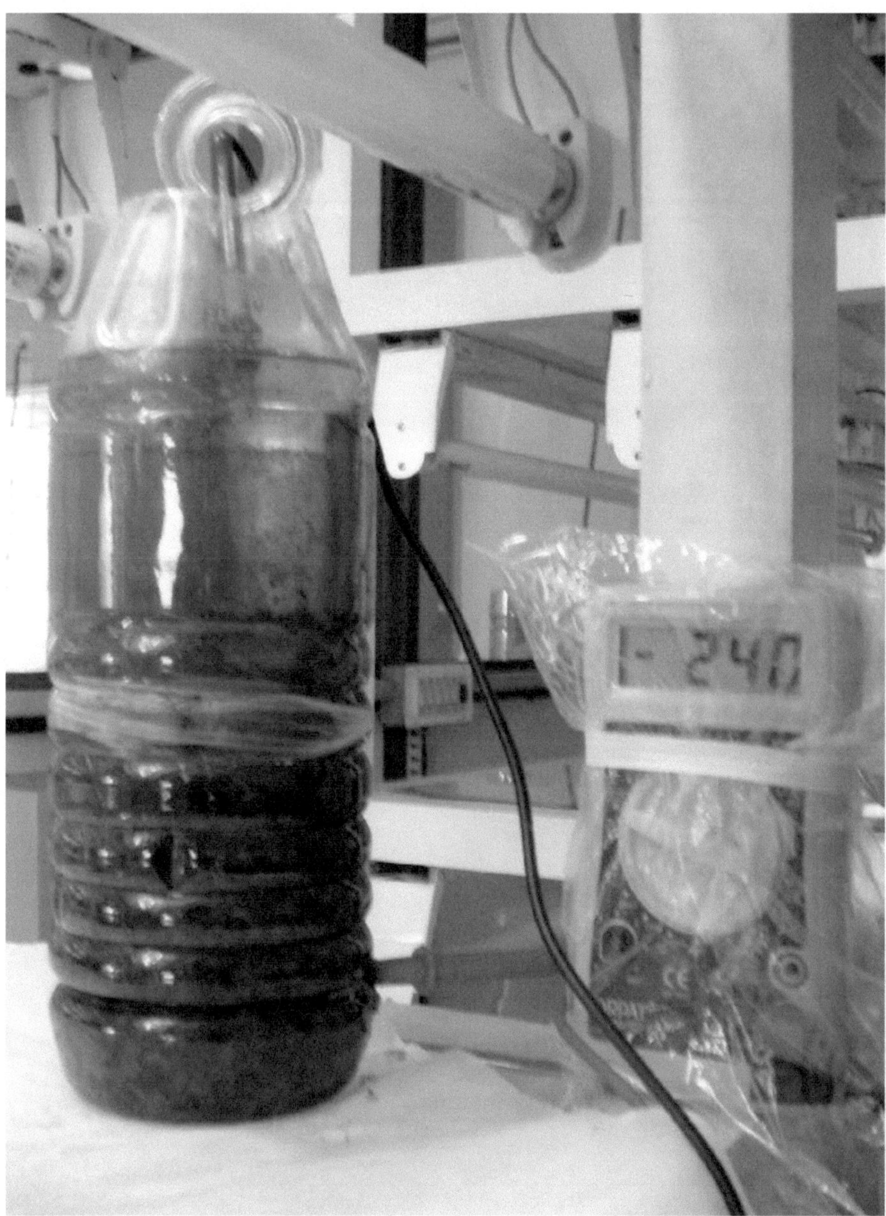

Figure 2

super capacitor 100 000mf 5.5V
+
Led diode
charge it with dc power
resistor from 220ohm to 470ohm

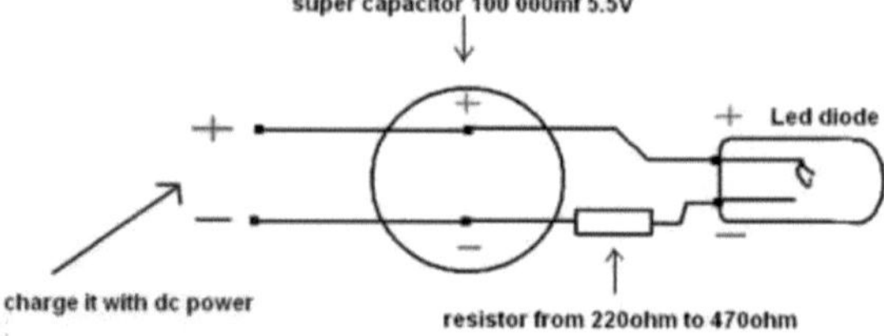

Figure 3

Figure 4

Discussion

This unique set up to harness redox electricity from Winogradsky column was efficient and kept giving power for more than a year till the biomass at the bottom was quite exhausted. This could be made into a custom made home decoration or a useful tool to light a LED bulb. All we need to do is just keep pouring water at the top to keep the Winograsky battery from running dry.

References

1. Anderson et al 1999

2. National Science Foundation, Microbe Power- Using a winogradsky column as a Battery, Lesson #6

3. Woodrow Wilson Summer Biology Institute Biodiversity, (2000), Investigating Bacteria with the Winogradsky Column

4. Super Cap Led 100 000 micro farads (www.instructables.com/Super Cap Led 100 000 micro farads, macobt

5. Sunil A. Patila, Falk Harnischa, Christin Kochb, Thomas Hübschmannc, Ingo Fetzerc, Alessandro A. Carmona-Martíneza, Susann Müllerc, Uwe Schröder, Electroactive mixed culture derived biofilms in microbial bioelectrochemical systems: The role of pH on biofilm formation, performance and composition, Bioresource Technology Volume 102, Issue 20, October 2011, Pages 9683-9690

6. http://My Winogradsky Column

7. http://instructable.com/supercapacitor